The Age of Technology:

19th Century American Inventors

edited and introduced by David C. King

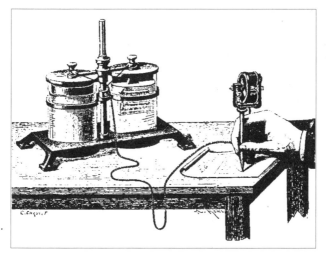

*Patented in 1877, Thomas Edison's electric pen was
used to create stencils for multiple copies of written
documents, before the typewriter was invented.*

Discovery Enterprises, Ltd.
Carlisle, Massachusetts

© Discovery Enterprises, Ltd., Carlisle, MA 1997

ISBN 1-878668-64-1 paperback edition
Library of Congress Catalog Card Number 96-84749

10   9   8   7   6   5   4   3   2   1

Printed in the United States of America

Subject Reference Guide:

Age of Technology:
19th Century American Inventors
edited and introduced by David C. King

Inventors — U. S. History
Technology — U. S. History

Credits:

Cover illustration (Clockwise from upper right):
Ford's Quadricycle, workings of a clock, the electric pen,
1867 steam-powered road roller, a sketch of the Hunley submarine,
and Howe's sewing machine.

Photos and illustrations courtesy of the National Archives,
except where noted in the text.

Table of Contents

Dedication

*To my family — Chris and Win, Mary and Harry —
for opening up new frontiers for research and writing.*

The American invents as the Greek
sculpted as the Italian painted. It is genius.

The Times *of London, 1876*

Introduction
by
David C. King

Late in the 19th century, Mark Twain wrote a novel about a Connecticut mechanic magically transported back to the days of King Arthur and the Knights of the Roundtable. The Yankee boasts about his inventive skills back home in 19th-century America:

> "Why I could make anything a body wanted — it didn't make any difference what; and if there wasn't any quick new fangled way to make a thing, I could invent one — and do it as easy as falling off a log...."

Mark Twain
A Connecticut Yankee in King Arthur's Court, 1889

The inventive spirit of Twain's Connecticut Yankee characterized the 19th century. The inventors who embodied that spirit transformed the world, laying the foundations for the abundance, comfort, and convenience we enjoy today. When Twain was born (as Samuel Longhorne Clemens) in 1835, the steamboat was the new and glittering marvel of American invention; the railroad was still in its infancy. By the time of Mark Twain's death in 1910, Americans were talking on telephones, lighting their cities and homes with electric lights, watching motion pictures, traveling across the country in comfortable railroad cars, and even bouncing around in the "new fangled" automobiles. They had also become the leaders of the world in mass producing food and manufactured goods.

After a slow beginning early in the century, the heroic age of invention really blossomed in the years after the Civil War. Between 1870 and 1900, the U. S. Patent Office issued 640,000 new patents. While the patents included such wonders as the telephone and electric light, most of the inventors of these patents made no fortunes and gained no lasting fame. But they kept right on inventing, motivated by that American notion that there must be a faster or better way of doing everything.

This book offers a brief survey of some of the highlights of that amazing era. There is not much techincal analysis of how each invention worked. Instead, we've tried to select readings that will bring out some of the drama of each story and indicate something of each invention's impact on our changing way of life.

Eli Whitney's cotton gin

The American System of Mass Production

Until the 19th-century age of invention, everything people used was made by hand. A clockmaker, for example, might spend several weeks fashioning a single clock. If a part of the clock later broke or wore out, the craftsperson had to fashion a new part that would fit with the others.

Almost everything we use today depends on mass production — the use of machines to turn out standardized parts at high speed and low cost. Our modern standard of living depends on this system; it would be impossible to return to the days of handcrafting the things we need and want.

The idea of replacing handcrafting with machines developed slowly. It became known as the "American System." The first steps in creating this system came from the inventive genius of Eli Whitney (1765-1825).

Eli Whitney: One Man, Two Revolutions

After graduating from Yale University in 1792, Whitney went south, planning to become a teacher. He was staying at the Georgia plantation of Mrs. Catherine Greene, widow of revolutionary war hero Nathaniel Greene, when he ran into a problem that was to change both his life and the history of the nation.

The problem was to find a way to make cotton a profitable crop. The only cotton that grew inland from the coast had sticky seeds that clung so stubbornly to the fiber that it took a worker ten hours to clean a single pound. In a letter to his father, Whitney described how he invented the cotton gin, a machine that was to revolutionize cotton production:

Source: *American Historical Review*, 1897-98, Vol. III, pp. 91-92.

...I heard much said of the extreme difficulty of ginning cotton, that is, separating it from its seeds. There were a number of very respectable gentlemen at Mrs. Greene's who all agreed that if a machine could be invented which would clean the cotton with expedition, it would be a great thing both to the country and to the inventor. I involuntarily happened to be thinking on the subject and struck out a plan of a machine in my mind....

In about ten days I made a little model, for which I was offered, if I would give up all right and title to it, 100 guineas. I concluded to relinquish my school and turn my attention to perfecting the machine. I made one before I came away which required the labor of one man to turn it and with which one man will clean ten times as much cotton as he can in any other way before known, and also cleanse it much better than in the usual mode. This machine may be turned by water or with a horse with the greatest ease, and one man and a horse will do more than fifty men with the old machines. It makes the labor fifty times less, without throwing any class of people out of business....

How advantageous this business will eventually prove to me, I cannot say. It is generally said by those who know anything about it that I shall make my fortune by it. I have no expectation that I shall make an independent fortune by it, but I think I had better pursue it than any other business into which I can enter. Something which cannot be foreseen may frustrate my expectations and defeat my plan....I wish you, sir, not to show this letter nor communicate anything of its contents to anybody except my brothers and sister, *enjoining* it on them to keep the whole a *profound secret.*

The invention had the simplicity of genius. A box contained a cylinder with dozens of bent wires. The wires worked in slots wide enough for cotton fiber to pass through, but not the seeds. By turning a crank, hooks pulled the cotton through the slots, then a revolving wire brush completed the cleaning.

Whitney obtained a patent and set up a factory for producing cotton gins in New Haven, Connecticut, but he never made his fortune from it. His early model was stolen and easily copied by others. Years of lawsuits followed and what money he did make was largely used up by lawyers' fees. His problems multiplied in 1795 when his factory burned to the ground. He wrote of his miseries to fellow inventor Robert Fulton in 1811:

Source: C.M. Green, *Eli Whitney and the Birth of American Technology* (Boston: Little, Brown Co., 1956), p. 77.

...I have labored hard against the strong current of Disappointment which has been threatening to carry us down the Cataract of Destruction — but I have labored with a shattered oar and struggled in vain.

In spite of the inventor's problems, cotton production soared from 4,000 bales in 1791 to 178,000 bales in 1810 and, by 1860, to more than 4,000,000 bales. The cotton fed mills in England and New England, where water-powered looms, invented in England, could turn out endless yards of cotton fabric. By the 1840s, 1200 American mills employed 70,000 workers.

Unwittingly, the cotton gin made the issue of slavery far more serious than ever before. The demand for cotton led to a sudden new demand for slave labor. The number of people held in bondage increased from less than one million in 1790 to more than four million in 1860, propelling the nation into the tragedy of Civil War.

Whitney's second invention was to have an equally powerful impact on history. In 1798, he wrote to the Secretary of the Treasury:

Source: J. Mirsky and A. Nevins, *The World of Eli Whitney* (New York: Collier Books, 1962).

I have a number of workmen and apprentices whom I have instructed in working woods and metals and whom I wish to keep employed....These circumstances induced me to address you and ask the privilege of having an opportunity of contracting for the supply of some articles the U. S. might want. I should like to undertake the Manufacture of ten or fifteen thousand stand of arms [muskets]....

I am persuaded that Machinery moved by water adapted to the business would greatly diminish the labor and facilitate the manufacture of this article. Machines for forging, rolling, boring, grinding, polishing, etc. may all be used to advantage.

Whitney knew nothing about making guns. But he had hit on a critically important idea — the idea of interchangeable parts. This is one of the essential ingredients of mass production, along with a power source and an assembly line. In 1801, hoping for more time to complete his machines, Whitney gave a demonstration of the gun locks to President John Adams and his Cabinet. He spread parts for the locks on the table and invited the men to assemble locks from any of the parts. Vice President Thomas Jefferson expressed his astonishment:

He had invented moulds and machines for making all the pieces of his lock so exactly equal that, take 100 locks to pieces and mingle their parts, and the hundred locks may be put together as well by taking the first pieces that come to hand. Good locks may be put together without employing a Smith.

Whitney continued to have trouble creating the machinery needed and he took ten years to complete an order he had promised in two years. But Whitney's Interchangeable — or American — System had been born. Slowly, the idea of standardized parts was applied to other products, with amazing results. When it was used for producing clocks, production soared and the price plummeted. By the 1840s, a single company had already sold more than 40,000 clocks overseas — a volume that would have been impossible a few years earlier.

Samuel Colt's Assembly Line

The next step in making mass production possible was the assembly line — having one worker complete one action, then pass the item along to the next person. The man most responsible for developing the assembly line was Samuel Colt (1814-1862), the man who invented the revolver, or six-shooter.

As a teenager in a small Massachusetts town, Sam was in trouble so often for experimenting with explosives that his father sent him to sea. On a voyage to India, Sam came up with the idea of making a pistol that would fire more than one shot before reloading. He carved a model of his "revolver" out of wood and, when he returned home, was determined to go into business. To raise money, he toured the country as "Dr. Coult," giving often hilarious demonstrations of nitrous oxide (laughing gas). By 1836, he had a patent and enough money to go into production.

Colt's revolvers and revolving rifles had a five or six-bullet chamber that revolved to the barrel and locked in place. After producing 5,000 guns, Colt's company went bankrupt when his financial backers cheated him. It was not until ten years later that he found out how valuable his invention was when he received a letter from Captain Samuel H. Walker of the Texas Rangers:

Source: W.B. Edwards, *The Story of Colt's Revolver* (Harrisburg, PA: Stackpole Co., 1953), p. 143.

Dear Sir,

The pistols you made for the Texas Navy have been in use by the Rangers for three years, and I can say with confidence that it is the only good improvement I have seen. The Texans have learned their value by practical experience, their confidence in them is so unbounded...that they are willing to engage four times their number. In the Summer of 1844, Col. J.C. Hays with 15 men fought about 80 Comanche Indians, boldly attacking them upon their own ground, killing and wounding about half their number. Several other skirmishes have been equally satisfactory.

Walker ordered 1,000 revolvers and other orders soon followed. Since he no longer had a factory, Colt turned to the Whitney Arms Company, now run by Eli Whitney, Jr. As profits grew, Colt built his own factory in Hartford, where he perfected the assembly line combined with Whitney's idea of interchangeable parts. In a letter to his father, he described how his system worked, with more than 400 machines in constant operation:

Source: Jack Rohan, *Yankee Arms Maker: The Incredible Career of Samuel Colt* (New York: Harper, 1935).

The first workman would receive two or three of the most important parts, and would affix these together and pass them on to the next who would do the same, and so on until the complete arm is put together. It would then be inspected and given the finishing touches by experts and each arm would be exactly alike and all of its parts would be the same. The workmen, by constant practice in a single operation would become highly skilled and at the same time very quick and expert at their particular task, so you have better guns and more of them for less money than if you hire men and have each one make the entire arm....

Colt made a huge fortune and his "six-shooter" became the standard weapon of soldiers, cowboys, outlaws and, later, 20th century lawmen and gangsters. The combination of Whitney's interchangeable system and Colt's assembly line was soon applied to manufacturing dozens of products. Between 1850 and 1900, the American System rapidly made the United States the world's leading producer of material goods. The age of abundance had arrived.

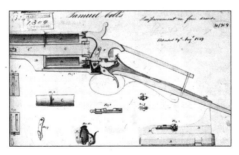

The Colt 45, U.S. Patent Office

Shrinking Time and Space

We have lived so long with high-speed travel and instant communication that it's hard for us to imagine just how slow transportation and communication were in the early 1800s. In January, 1815, for example, Americans won a spectacular victory at the Battle of New Orleans, the final battle of the War of 1812; but the battle was fought more than two weeks after a peace treaty, ending the war, had been signed in Europe. It had taken that long for news of the treaty to cross the Atlantic Ocean.

The speeding up of both traveling and sending messages depended on harnessing two new sources of power — steam and electricity. The steam engine was invented in England in the 1700s to pump water out of coal mines. Knowledge of electricity and the creation of storage batteries were developed by several scientists and inventors, with Benjamin Franklin one of the earliest.

"Mr. Fulton's Folly"

The steam engine was applied first to steamboats and then to railroads. Robert Fulton (1765-1815) did not invent the steamboat. A dozen European inventors had experimented with steam-powered boats in the late 1700s. An American inventor, John Fitch (1743-1798) operated a steamboat on the Delaware River from 1787 until 1791 and called himself the "Lord High Admiral of the Delaware." But the public was not ready for steamboat travel, the venture failed, and Fitch died penniless.

After a rather mediocre career as an artist, Connecticut-born Robert

Fulton traveled to Europe and turned his attention to inventing. He developed ideas for improving canals and successfully developed both a submarine and torpedoes. When these experiments failed to arouse much interest, he decided to try a steamboat. In 1806, he designed the *North River Steamboat of Clermont*, using a British-made steam engine to turn two side paddle wheels. In September, the *Clermont* was ready for its maiden voyage up the Hudson River from New York to Albany. Here is how Fulton described the historic undertaking:

Source: Quoted in Emily Davie, ed., *Profile of America: An Autobiography of the USA*, (New York: Thomas Y. Crowell Co., 1954), pp. 199-200.

When I was about to build my first steamboat, the public of New York in part regarded it with indifference, in part with contempt, as an entirely foolish undertaking. My friends were polite, but they were shy of me....[Others] laughed aloud, and made jokes at my expense; and reckoned up the fallacy and loss of money on "Mr. Fulton's Folly."

At length came the day when the experiment was to be tried. To me it was a moment of the utmost importance. I had invited many of my friends to go on board and witness the first successful voyage. Many of these did so reluctantly, and in the belief that they should become the witnesses of my humiliation, and not of my triumph; and I know very well there was sufficient reason to doubt my success. The machinery was new and ill made. A great portion of it was prepared by artisans unaccustomed to such work; and difficulties might easily arise, also, from other causes. The hour arrived at which the boat was to begin to move. My friends stood in groups on deck. Their looks indicated uneasiness, mixed with fear; they were silent and dejected. The signal was given, and the boat was put in motion; it advanced a short distance and then stopped, and became immovable. The former silence now gave way to murmurs, and displeasure, and disquiet whisperings and shrugs of shoulders. I heard on all sides "I said it would be so"; "It is a foolish undertaking"; "I wish we were well out of it."

I mounted the platform and told my friends that I did not know what was the cause of the stoppage, but if they would be calm, and give me half an hour's time, I would either continue the voyage or give it up entirely. I went down to the engine, and very soon discovered an unimportant oversight in the arrangement: this was put to rights. The boat began to move once more. We left New York; we passed through the Highlands; we arrived in Albany!

The *Clermont* completed the 300-mile round trip in a little over 60 hours, averaging five miles per hour. Not exactly blinding speed, but it marked the true beginning of steam-powered travel. Fulton went on to build twenty more steamboats and enjoyed both fame and financial success. As he predicted, the greatest use of the steamboat would be on the western waters of the Mississippi River system. By the 1840s, dozens of steamboats were carrying ten million tons of freight and thousands of passengers on the Misssissippi. As one newspaper declared, "...steam — the tiny thread that sings from the spout of a tea-kettle ...suddenly steps forth...and annihilates time and space." (Source: New York *Mirror*, Nov. 28, 1840.)

Steam provided power to turn the boat's paddle wheels.

The Railroad

The steam-powered railroad was an extension of the steamboat to land travel. The first railroads were actually used for carriages drawn by horses. English inventors first applied the idea of using a steam engine for power, and Americans followed this lead in the early 1830s. Even earlier, in 1813, Oliver Evans, one of the unheralded pioneers of American invention, predicted:

Source: G. & D. Bathe, *Oliver Evans: A Chronicle of Early American Engineering* (PA Historical Society, 1935; reprinted by Arno Press, 1972).

The time will come when people will travel in [stage coaches] moved by steam engines, from one city to another, almost as fast as birds can fly, fifteen to twenty miles an hour....Two sets of rails will be laid, with a rail to guide the carriage, so that they may pass each other in different directions and travel by night as well as by day....

This 1867 steam-powered road roller
paved the way for the motor car.

The first American railroads began operating in the early 1830s. Some people worried about the great speed of "fifteen to twenty miles an hour," and most found them anything but comfortable. As English author Charles Dickens observed, "There is a great deal of jolting, a great deal of noise, a great deal of wall, not much window, a locomotive engine, a shriek, and a bell." (Source: *American Notes*, New York: Penguin Books, 1972.)

The American public soon took to railroading and, by the 1850s, more than 30,000 miles of track were knitting the nation together. A number of American inventors made important improvements. In the 1840s, for example, Robert Livingston Stevens developed the T-shaped rail, the spike, and wooden ties on a gravel bed — all still standard today. Improvements in safety and comfort came later.

Morse and the Telegraph

Samuel F. B. Morse (1791-1872) was an artist, not an inventor. He knew almost nothing about electricity. And yet his invention of the telegraph had a more instantaneous impact on the shrinking of time and space than any other single invention.

Born in Massachusetts, Morse struggled for years as a portrait painter and landscape artist. Modern art historians consider him a first-rate talent, but he found he could not make a living at it. On a voyage from England in 1832, a conversation with fellow passengers led him to start working on the telegraph — using electricity to transmit messages. It marked the beginning of more than ten years of struggle.

Morse made a simple model consisting of a battery, two sets of wires for transmission in each direction, and a mechanism for breaking and restoring the circuit at one end; at the other end, a receiving device held a pencil for recording the impulses on a moving strip of paper. He built the model out of a frame used to stretch an artist's canvas. Joseph Henry (1797-1878), a brilliant scientist, had already discovered the same ideas and helped Morse with advice. In fact, if Henry had bothered to patent his own ideas, he would have been credited with inventing the telegraph. The main new thing Morse contributed was his Morse Code, using a series of dots and dashes for each letter in the alphabet.

Morse was ready to try his device, but he needed funds and turned to the government. In 1838, a Congressional report was enthusiastic:

Source: E.L. Morse, ed., *Samuel F.B. Morse: His Letters and Journals* (Boston: Houghton Mifflin, 1914), p. 186.

It is obvious that the influence of this invention…will, in the event of success, of itself amount to a revolution unsurpassed in moral grandeur by any discovery that has been made in the arts and sciences.…Space will be, to all practical purposes of information, completely annihilated between the States of the Union.

Glowing praise — but no money. Not until 1843 did Congress provide $30,000 to construct a 30-mile telegraph line between Washington, D.C. and Baltimore. On May 24, 1844, Morse tapped out the historic first message, from Baltimore to the Supreme Court Building in Washington: "What hath God wrought!"

The result was a revolution in communication. Telegraph lines soon criss-crossed the nation. Messages that had once taken days or weeks to deliver, could now be transmitted in a matter of seconds. Perhaps the fastest means of communication before the telegraph became widely used was the Pony Express, organized in 1860, which could deliver mail 2,000 miles between St. Louis and San Francisco in just ten days. Morse's invention reduced the time to less than ten minutes.

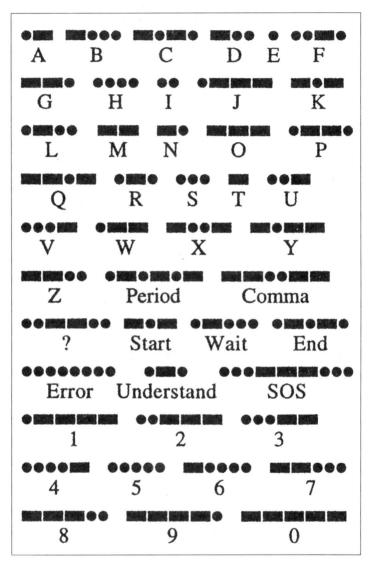

The alphabet, numbers, and commands of the Morse Code, used to transmit telegraph messages by a series of dots and dashes

Although Morse eventually gained both fame and a considerable fortune, he first had to wage battles in court to uphold his patent. He finally won every lawsuit, but he was pained by the attacks on him:

Source: E.L. Morse, ed., *ibid.*, p. 224.

I am held up by name to the odium of the public....I find the fate of Whitney before me....Take as much pains as you will to secure yourself, and your valuable invention, you will be robbed of it and abused in the bargain. This is the lot of the successful inventor and no precaution, I believe, will save him from it....the unprincipled will hate him and detract from his reputation to compass their own contemptible and selfish ends.

The Crystal Palace

In 1851, what can be described as the first modern world's fair was held in London. Called the Great Exhibition at the Crystal Palace, the event offered all nations an opportunity to display their arts, crafts, and manufactured goods. Visitors to the exhibits were so surprised by the number and variety of American-made goods that the British government sent two commissioners to the United States to find out how the Americans were producing so much. The commissioners reported:

Source: Quoted in Bruce Norman, *The Inventing of America* (New York: Taplinger Publishing Co., 1976), p. 121.

The American method of Manufacturing Principle is the production of large numbers of standardized articles produced on the basis of repetition in factories — whereas in England handicrafts and work outside the factory still persist....[We] could not fail but be impressed from all that we saw there, with the extraordinary energy of the people, and their peculiar aptitude in availing themselves to the utmost of the immense natural resources of their country.

The British, who had long prided themselves on being the world's most technologically advanced people, were beginning to realize that the Americans were forging ahead of them. And the Yankee enterprise of inventions and machines was just beginning. The pace of invention indicates what was in store: The U.S. Patent Office issued 5, 942 patents during the 1840s, and 23,140 in the 1850s. Between 1870 and 1900, the government was issuing more than 20,000 patents every year.

Elisha Otis: Conquering Vertical Space

Vermont-born Elisha Otis (1811-1861) was working as a construction supervisor in New York City in 1852 when he came up with the idea for a safe mechanical elevator. A few elevators were in existence to carry freight four or five stories — as high as buildings were constructed in the 1850s. They were held up by a counterweight that had to go as deep into the ground as the building was high. The ropes sometimes frayed and broke, which made the elevators unsuitable for people.

Otis invented a safety device which he demonstrated at New York's version of the Crystal Palace in 1854; after being hoisted to the top of the display, he dramatically ordered that the rope be cut. When the elevator didn't fall, his point was made. He formed the Otis Elevator Company and his sons later added electrical power to the elevators. This single invention made possible a revolution in architecture and construction. Taller buildings were now possible, and the first skyscrapers were soon under construction.

A visitor to Chicago described the sensation of this vertical transportation:

Source: Quoted in Bessie Louise Pierce, ed., *As Others See Chicago* (Chicago, 1933).

"The slow-going stranger...feels himself loaded into one of those...baskets...and the next instant up goes the whole load as a feather caught up by a gale. The descent is more simple. Something lets go, and you fall from ten to twenty stories as it happens."

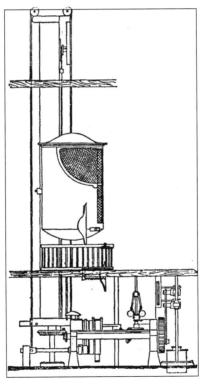

William Ellery Hale made a fortune with his early passenger elevator.

Revolutions in Farming and Food Production

In the early years of the 19th century, farmers were not interested in changing their time-honored tools and methods, and many were suspicious of new technology. For example, when a plow with an iron-covered blade was introduced in the 1790s, American farmers would not use it, saying that the iron would poison the soil.

By the 1830s, farm families realized that new tools could increase production and profits. In the 1830s, John Deere (1804-1886) developed a plow with a steel blade. The steel made it possible for the first time to plow America's Great Plains, a vast area that had been known as the "Great American Desert," because it was believed that a plow could not cut through the thick sod.

With Deere's plow, farmers could now turn the entire prairie into wheat fields. First, however, they needed a way to harvest what they planted. Wheat ripens all at once; what cannot be harvested in the space of about two weeks is lost. A prairie farmer described the problem in 1839:

Source: Quoted in Marshall B. Davidson, *Life in America*, Vol. 1, (Boston: Houghton Mifflin Company, 1951), p. 411.

"You can behold the vast plain of 12,000 acres, all waving in golden color, ripe for the cradle. At this moment, every man and boy, and even women, are actively engaged in cradling, raking, binding, and shocking the golden harvest. ...But, after all, a large portion will be left out, and will be destroyed. There is not *help* enough in the country to secure the crop."

The "help" came, not in the form of more workers, but in mechanizing the harvest. The pioneer in this agricultural revolution was Cyrus H. McCormick (1809-1884).

The McCormick Reaper

In 1831, on his family's farm in Virginia, McCormick worked out the basic principles of a machine to harvest wheat, solving problems that had stumped other inventors. The horse-drawn reaper, for example, had a reciprocating blade — that is a blade that cut backward and forward, so that *all* the stalks were cut. The blade also had teeth which held the stalks upright, and a curved arm, or divider, bent the wheat toward the blade. A heavy wheel, located directly behind the horse, provided the traction to turn the machinery. The cutter was located on the side, so that the horse would walk only on stubble already cut. Yet, the cut grain still had to be raked and bound; machines for that were later added to the reaper making it a "combine."

While McCormick's Virginia neighbors saw little advantage in using a reaper on their small, hilly lands, the inventor found his market on the prairie. He opened a factory in Chicago and advertised the reaper by holding field trials. The importance of those demonstrations was reflected in the response to the reaper at London's Crystal Palace:

Source: The *Times* of London, 1851, quoted in Davidson, *ibid.*, p. 412.

On first viewing the reaper, the *Times* of London described it as

"...a cross between a flying machine, a wheelbarrow and an Astley chariot [a carriage]...an extravagant Yankee contrivance, huge, unwieldy, unsightly and incomprehensible."

After a field trial showed the reaper's quiet speed, the same newspaper concluded:

> "The reaping machine from the United States is the most valuable contribution from abroad to the stock of our previous knowledge....It is worth the whole cost of the Exposition."

American newspapers also extolled the virtues of McCormick's machine. An example:

Source: *Philadelphia Photographer*, August, 1864.

> "The saucy machine has driven the scythe from the field, almost tempting old Time to choose a new weapon. Now the principal work of the harvest is to drive the horse about the field a few times and lo! the harvest is gathered."

Although farming was not *quite* as easy as driving a reaper around the field "a few times," the reaper did lead to an enormous growth in production. By 1860, more than 80,000 reapers were cutting oceans of grain in America's new "breadbasket" and the nation's wheat harvest doubled and doubled again.

Swift and Armour: Preserving Food

The tremendous increase in food production went hand-in-hand with the growth of cities and the expansion of industry. In 1800, nearly 90 percent of the nation's families lived by farming; by 1900, a farm family produced enough to meet its own needs and the needs of several city dwellers, so that barely half the population now lived on farms. An important step in feeding the growing cities was in packaging the food and getting it to markets.

Two pioneers in this field were Gustavus F. Swift (1839-1903) and Philip Danfourth Armour (1832-1901). In 1875, Swift moved to Chicago and came up with the idea of shipping dressed beef to eastern cities, thus avoiding the expense of shipping live cattle. He designed a refrigerated railroad car, and developed very precise methods for keeping the meat fresh from the packing house to the customer. The plan was a great success and, by the 1890s, Swift and Company was operating in 400 cities worldwide, with more than 30,000 workers. Armour followed a parallel path, also using Chicago as a base, and developing pork products as well as beef. In the early 1900s, the Armour Company passed Swift as the world's leading meat packer.

Gail Borden's Condensed Milk

Gail Borden was one of many inventors who found new ways to package food, including the use of tin cans. (Canning was developed in France early in the 1800s and later improved by several English inventors.) Borden pursued the idea of condensing food. He first came up with a biscuit made of dried meat. He managed to win a prize at London's Crystal Palace in 1851, but the Meat Biscuit was a commercial flop, partly because the taste and texture were so awful.

Borden persevered and turned his attention next to condensing milk. He borrowed the ideas of a Shaker community that was using a vacuum evaporator in processing maple syrup. Borden tinkered with the system and, after months of trial and error, had a process that he patented. An advertisement in 1858 presented Borden's claim:

Source: *Frank Leslie's Illustrated Newspaper*, May 22, 1858.

Borden's Condensed Milk, Prepared in Litchfield County, Conn., is the only Milk ever concentrated without the admixture of sugar or some other substance, and remaining easily soluble in water. It is simply Fresh Country Milk, from which the water is nearly all evaporated, and nothing added. The Committee of the Academy of Medicine recommend it as "an article that, for purity, durability and economy, is hitherto unequalled in the annals of the Milk Trade."

One quart, by the addition of water, makes 2 1/2 quarts of cream, or 5 quarts rich milk, and 7 quarts good milk. For sale at 173 Canal Street, or delivered at dwellings in New York and Brooklyn, at 25 CENTS per quart.

Borden's condensed milk helped convince people that canned foods not only preserved food, but were also quick and easy to use. During the Civil War, the Union Army found that the small Borden cans were perfect for soldiers' field rations. As the soldiers returned home, they helped spread the popularity of Borden's product and of canned foods in general. By 1870, American companies were producing 30 million canned foods each year.

Instant Breakfast Food

Dr. John Harvey Kellogg (1852-1943) was a pioneer in persuading Americans to use ready-to-eat foods. Breakfast, for example, had always been a heavy meal that took time to prepare — oatmeal, eggs, bacon, and potatoes. In 1893, Dr. Kellogg, the director of a health sanitarium in Battle Creek, Michigan, invented corn flakes. This is how he told the story:

Source: Gerald Carson, "Cornflake Crusade," *American Heritage*, June, 1957, pp. 66, 85.

I prescribed zwieback [a hard bread that has been sliced and rebaked] for an old lady, and she broke her false teeth on it. She demanded that I pay her ten dollars for her false teeth. I began to think we ought to have a ready-cooked food which would not break people's teeth. I puzzled over that a good deal.

One night about three o'clock, I was awakened by a phone call from a patient, and as I went back to bed I remembered that I had been having a most important dream. Before I went back to sleep again, I gathered up the threads of my dream, and found I had been dreaming of a way to make flaked foods.

The next morning I boiled some wheat, and while it was soft, I ran it through a machine Mrs. Kellogg had for rolling dough thin. This made the wheat into thin films, and I scraped it off with a case knife and baked it in the oven. That was the first of the modern breakfast foods.

Dr. Kellogg and his brother, Will Keith Kellogg, soon found that corn flakes were much more palatable, and Will started the Kellogg Company. Rivals soon engaged them in the "cereal wars" for the most salable product, a competition that helped popularize ready-to-eat breakfast foods.

Inventing the Modern Age

The first three quarters of the 19th century had produced wonders of invention and production. But that was only the beginning. The last quarter of the century was a magic-carpet ride into a modern civilization that people like Whitney and Fulton would not have recognized. Historian Bernard A. Weisberger describes the achievements of the inventors between 1875 and 1900:

Source: Bernard A. Weisberger, *The LIFE History of the U.S., Vol. 7, The Age of Steel and Steam* (NY: Time, Inc., 1964), p. 38.

They set the gloom of night aglow with cheery electric light, enabled the people in all corners of the land to talk to one another. They made machines that talked, cameras a child could work, typewriters, barbed wire and workable safety pins. As they produced their marvels, the soldiers and statesmen, the nation's heroes for two-thirds of the century, faded away. Now the heroes were the inventors — Thomas Alva Edison, Alexander Graham Bell, John Augustus Roebling, George Eastman, George Westinghouse....They built a new civilization, based on machines and mass production.

"The Sewing-Machine War"

An 1846 Howe sewing machine

The story of the invention of the sewing machine began in the 1840s, but more than twenty years were needed to perfect it and to settle the war over patent rights. The machine was widely advertised as a boon to homemakers, speeding and simplifying the task of making clothes. But its greatest importance was in making possible the mass production of ready-made clothes. Money was no longer necessary for dressing well; now everyone, as one advertisement said, "can be as well clad as any millionaire."

Several inventors had tried to produce a machine that would sew. The first to succeed was Elias Howe (1819-1867). In the 1840s, he created a machine with two needles, each with an eye in the tip; the upper and lower needles worked together to produce 250 even stitches per minute, seven times faster than by hand. Like so many inventors before him,

Howe found no buyers, partly because the price of $300 was too steep for manufacturers.

After obtaining a patent in 1846, Howe made an unsuccessful trip to England. He returned in 1849 in poverty, his household goods were lost in a shipwreck, and his wife died soon after their arrival. Adding to his misery was the discovery that several rivals were manufacturing sewing machines, using his ideas.

Howe's main rival was Isaac M. Singer (1811-1875). Born on a farm in upstate New York, Singer had run away from home to start a career as actor, mechanic, and inventor. In a matter of weeks, he had made important improvements in Howe's machine:

Source: *Genius Rewarded: The Story of the Sewing Machine* (New York: Singer Sewing Machine Co., 1880).

Instead of the shuttle going around in a circle, I would have it moved to and fro in a straight line and in place of the needle bar pushing a curved needle horizontally, I would have a straight needle and make it work up and down.

Singer's improvements created a machine that could sew different kinds of seams and could be operated by a foot treadle rather than one hand. The inventions of both men were necessary for a machine that could mass produce clothing.

Howe sued Singer and his other rivals and eventually won in court, earning a royalty for Howe on every machine sold during the length of his patent. While Howe made his fortune, it was Singer who developed one of the world's giant corporations. He was one of the first to spend heavily on advertising and it paid off. By the 1870s, the Singer Company was selling 500,000 sewing machines a year. Advertising continued to emphasize the blessings of the sewing machine for the homemaker:

Source: Quoted in Davidson, *op. cit.*, p. 513.

When women toiled for daily bread from
  early morn to eve,
How many eyes were dimmed with tears,
  how many hearts did grieve;
But now she has her "household pet"
  and one to which she'll cling.
For labor is a pleasure now, and she
  can toil and sing —
It works alike for rich and poor, the
  humble and the proud.

Bell's Telephone: The Most Valuable Patent in History

In 1876, the United States celebrated the nation's Centennial with a magnificent Centennial Exhibition in Philadelphia. One of the smallest displays in any of the buildings was a table containing Alexander Graham Bell's telephone, which the inventor had patented a few weeks before the Exhibition opened. Only a few of the eight million visitors paid much attention to the strange-looking device. Fewer still suspected how important this invention was to become.

Scottish-born Alexander Graham Bell (1847-1922) was, like his father, a teacher of the deaf. After migrating to Canada, then to the United States, he became increasingly interested in the idea of reproducing sound. The Western Union Telegraph Company, which by now had strung some 400,000 miles of telegraph wires, was looking for a way to send several telegraph messages over the same wire at the same time. The prize offered was a good incentive to young inventors like Bell. "If I can make the deaf talk," he declared, "I can make iron talk."

Bell described his ideas to Joseph Henry, the elderly scientist who had helped Samuel Morse thirty years earlier. In a letter to his parents, Bell explained:

Source: R.V. Bruce, *Alexander Graham Bell and the Conquest of Solitude* (London: Gollancz Co., 1973).

It would be possible to transmit sounds of ANY SORT if we could [create] a variation in the intensity of an electric current like that occurring in the density of air....

My inexperience in such matters is a great drawback. However, Morse conquered his electrical difficulties even though he was only a painter, and I have no intention of giving in either....

[Joseph Henry] said he thought it was "the germ of a great invention"....I said I recognized that there were mechanical difficulties....I added that I felt that I had not the electrical knowledge necessary to overcome the difficulties. His laconic answer was "Get it." I cannot tell you how much those two words encouraged me.

Bell poured through books and journals to learn enough about electricity. He also hired a young mechanic, Thomas A. Watson, to construct the devices he needed.

The equipment at first consisted of three telegraph transmitters and receivers, connected by wires and placed in different rooms. Each device had metal strips he called "reeds" that could vibrate to produce a sound. The vibrating reeds caused fluctuations in the electrical current that were reproduced on the receiving end. For a diaphragm to reproduce the voice sounds, he used a platinum needle dipped in acid. With this rudimentary device, he uttered the first telephone message, "Mr. Watson, come here, I want you." The system was far from perfect, but it was enough for him to obtain his patent and demonstrate it at the Centennial Exhibition, where the Emperor of Brazil exclaimed, "My God, it talks!"

Bell's first telephone

Reactions to what Bell called the tele-phone were mixed. Here is a sampling:

Source: New York *Tribune*, Nov. 4, 1876.

The Centennial Exhibition has afforded the opportunity to bring into public view many inventions and improvements which otherwise would only have been known to the smaller circles....The tele-phone is a curious device that might fairly find place in the magic of Arabian Tales. Of what use is such an invention? Well, there may be occasions of state when it is necessary for officials who are far apart to talk with each other without the interference of a [tele-graph] operator. Or some lover might wish to pop the question directly into the ear of a lady and hear for himself her reply, though miles away; it is not for us to guess how courtships will be carried on in the Twentieth Century....

Source: Scottish scientist Sir William Thomson, quoted in Bruce Norman, *ibid.*, p. 56.

With my ear pressed against a disc, I heard it speak distinctly several sentences...I need scarcely say I was astonished and delighted....With somewhat more advanced plans, and more powerful apparatus, we may confidently expect that Mr. Bell will give us the means of making voice and spoken words audible through the electric wire to an ear hundreds of miles distant.

Source: The *Times* of London, July, 1876.

The latest American humbug — far inferior to the well-established speaking tubes....

Western Union agreed with *The Times*. When Bell offered to sell his invention for $100,000, the company president turned him down, saying, "What use can we make of an electrical toy?"

Bell, with financial backing from his father-in-law, started his own company. Thomas Alva Edison contributed a key component to the improved telephone by inventing a transmitter using a carbon button, not much different from the transmitter on modern telephones. Western Union bought Edison's transmitter and hired him to compete with Bell, clearly infringing on Bell's patent, while the Bell Telephone Company freely used Edison's transmitter. Other inventors, including Elisha Gray and Emile Berliner, became involved and years of court wrangling followed — a total of some 600 lawsuits.

The legal battles didn't prevent Edison from gaining a profit and having fun, while Bell deservedly earned a considerable fortune and immense popularity. And the early derision of the telephone did not interfere with its amazing growth. By 1880, only four years after Bell's patent was granted, there were 50,000 telephones in operation. Within twenty years, there were more than one million. Bell, now a wealthy man at the age of thirty, left the business in 1881 and returned to teaching the deaf. He later worked on other inventions and became involved in the earliest attempts to create an airplane.

SPEAKING TELEPHONES.

THE AMERICAN BELL TELEPHONE COMPANY,

| W. H. FORBES, President. | W. R. DRIVER, Treasurer. | THEO. N. VAIL, Gen. Manager. |

Alexander Graham Bell's patent of March 7, 1876, owned by this company, covers every form of apparatus, including Microphones or Carbon Telephones, in which the voice of the speaker causes electric undulations corresponding to the words spoken, and which articulations produce similar articulate sounds at the receiver. The Commissioner of Patents and the U. S. Circuit Court have decided this to be the true meaning of his claim; the validity of the patent has been sustained in the Circuit on final hearing in a contested case, and many injunctions and final decrees have been obtained on them.

This company also owns and controls all the other telephonic inventions of Bell, Edison, Berliner, Gray, Blake, Phelps, Watson, and others.

(Descriptive catalogues forwarded on application.)

Telephones for Private Line, Club, and Social systems can be procured directly or through the authorized agents of the company.

All telephones obtained except from this company, or its authorized licensees, are infringements, and the makers, sellers, and users will be proceeded against.

Information furnished upon application.

Address all communications to the

AMERICAN BELL TELEPHONE COMPANY,
95 Milk Street, Boston, Mass.

Advertising blossomed with the arrival of new inventions.

Thomas Alva Edison: "The Wizard"

One of the least-noticed award winners at the Centennial Exhibition was a young man named Thomas Alva Edison (1847-1931). He won an award for an improvement on the telegraph. No one at the Exhibition could have guessed that this slouched figure in a rumpled suit would become the dominant personality in the field of invention for the next quarter century.

No one, that is, except Thomas Alva Edison himself. By 1876, he knew he had a genius for invention, and he was determined to astound the world with his feats. He succeeded. Edison was also a great self-promoter and loved the spotlight of publicity. Reporters were always willing to cover his exploits and his often-grandiose claims because he always made news. To the American people, he came to symbolize the ideal rags-to-riches story, and he also symbolized the lonely inventor, toiling through the night in search of some breakthrough. Partially deaf since childhood, Edison even managed to make that part of his mystique.

A quick summary of his achievements includes: (1) solving the telegraph problem of sending more than one message over the same wire; his Duplex Telegraph was quickly followed by the Quadruplex, relaying two messages going out and two coming in; (2) the carbon button transmitter for the telephone; (3) the phonograph; (4) the incandescent light bulb; (5) an entire system for generating and delivering electric power to homes and businesses; (6) an electric pen for making stencils; (7) the mimeograph machine; (8) an electric railroad; (9) waxed paper; (10) the basic apparatus for motion pictures.

Edison accomplished all of these before 1900. After the turn of the century, he enjoyed being the elder statesman of invention, a man who held the amazing record of 1,093 patents. (For more on Thomas Edison, see Discovery Enterprises's *Thomas Alva Edison: The King of Inventors*.)

Here are some highlights of Edison's extraordinary career:

The Phonograph

Edison described how working on the transmitter for Bell's telephone led him to the idea of a phonograph:

Source: W.K. Dickson, *The Life and Inventions of Thomas Alva Edison* (London: Chatto & Windus, 1894), p. 122.

I discovered the principle by the merest accident. I was singing to the mouthpiece of a telephone when the vibrations of the voice sent the fine steel point into my finger. That set me thinking. If I could record the actions of the point and send the point over some surface afterward, I saw no reason why the thing would not talk. I tried the experiment first on a string of telegraph paper [waxed strips about half an inch wide] and found that the point made an alphabet [i.e., marked the paper with tiny holes]. I shouted the words "Hallo! Hallo!" into the mouthpiece, ran the paper back over the steel point and heard a faint 'Hallo! Hallo!' in return. I determined to make a machine that would work accurately and gave my assistants instructions, telling them what I had discovered. That's the whole story. The phonograph is the result of the pricking of a finger.

Edison's phonograph consisted of a cylinder covered with tinfoil, with a simple needle and diaphragm, and another needle and diaphragm for playback. When it was made public, the instrument caused a sensation. Although the phonograph could only record and play back about a minute of sound, and that with poor quality, people flocked to "phonograph parlors" that sprang up around the country during 1878.

The "phonograph craze" died out quickly, once people's curiosity was satisfied. Edison set the invention aside, not realizing its enormous potential. Over the next ten years, several inventors, including Bell

The Edison phonograph

and Emile Berliner, made important improvements. Edison hastily rejoined the competition in 1888. After days of around-the-clock work, he had an improved phonograph. To promote his invention, Edison invited famous people to record their voices. One of the first was the famous English composer Sir Arthur Sullivan; his recording stated:

Source: Quoted in Ernest V. Heyn, *A Century of Wonders: 100 Years of Popular Science* (Garden City, NY, Doubleday, 1972), p. 90.

For myself, I can say only that I am astonished and some-what terrified at the results of this evening's experiments. Astonished at the wonderful power you have developed, and terrified at the thought that so much hideous and bad music may be put on record forever.

By 1900, the combined work of Edison, Bell, and Berliner produced machines of high quality, and Berliner's wax disc records soon replaced Edison's cylinders.

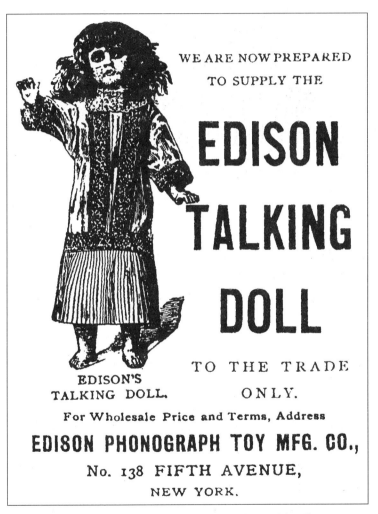

Spin-offs of popular new inventions were advertised heavily.

Electric Lighting

One reason Edison had stopped work on the phonograph was a sudden new passion — electric lights. Edison did not invent electric lighting. By 1878, several cities in Europe and the U.S. were partially illuminated by "arc lights" — a current of electricity flowing between two carbonized poles or rods. Arc lights used such a huge amount of electricity and produced such a brilliant light that they could only be used for street lighting. The problem was that the electrical current could not be subdivided into smaller units that could be used in homes and businesses. A British government report in 1878 stated emphatically:

> "The sub-division of the electric light is a problem that cannot be solved by the human brain."

To solve the problem, several inventors were trying to create a "glow lamp" — a lamp in which electric current caused a rod, or filament, to glow without burning up. This is *incandescent* light, and this was Edison's great triumph. This is how Edison described the problem:

Source: George Parsons Lathrop, "Talks With Edison,"*Harper's Magazine*, LXXX, Feb., 1909.

To the question: "Which invention caused you the most study? Thomas Edison replied:...The electric light. For, although I was never myself discouraged, or inclined to be hopeless of success, I cannot say the same for all of my associates. And yet, through all those years of experimenting and research, I never once made a discovery. All my work was deductive, and the results I achieved were those of invention pure and simple. I would construct a theory and work on its lines until I found it was untenable. Then it would be discarded at once, and another theory evolved. This was the only possible way for me to work out the problem, for the conditions under which the incandescent electric light exists are peculiar and unsatisfactory for close investigation. Just consider this: we have an almost infinitesmal filament heated to a degree which it is difficult for us to comprehend, and it is in a vacuum, under conditions of which we are wholly ignorant. You cannot use your eyes to help you in the investigation, and you really know nothing of what is going on in that tiny bulb. I speak without exaggeration when I say that I have constructed *three thousand* different theories in connection with the electric light, each one of them reasonable and apparently likely to be true. Yet in only two cases did my experiments prove the truth of my theory.

After months of effort, Edison found that carbonized bamboo created a long-lasting filament, and he had his incandescent light bulb. Bamboo remained the standard filament until the modern tungsten filament was developed a few years later.

Edison turned next to what was probably his greatest achievement — creating an entire system to deliver electric lighting to customers. This meant devising generators to produce the electrical current, an underground system of wiring to carry the current to homes and businesses, meters to measure the amount of electricity used, as well as switches for lamps and sockets to hold the bulbs. It took nearly three years to develop all these components and to build the first power station in New York City. Every step involved unforeseen difficulties, like Edison's account of a generator problem:

Source: George S. Bryan, *Edison: The Man and His Work* (New York: Alfred A. Knopf, 1926).

Of all the circuses since Adam was born, we had the worst then. One engine would stop and the other would run up to a thousand revolutions, and they would see-saw....When the circus commenced, the men who were standing around ran out precipitately, and some of them kept running for a block or two. I grabbed the throttle of one engine and E.H. Johnson, who was the only one present to keep his wits, caught hold of the other and we shut them off....

Even after the power station was opened, many observers remained convinced that electric lights would not replace gas light. As late as 1895, *Popular Science* concluded:

Source: *Popular Science*, quoted in Ernest V. Heyn, *Fire of Genius: Inventors of the Past Century* (NY: Doubleday, 1976), p. 135.

Not withstanding the rapid development of electric lighting, the use of gas in dwelling houses, offices, and stores is undoubtedly so convenient and comparatively safe that for many years it will remain the chief means of artificial illumination.

Even as that prediction was being made, hundreds of generating plants were under construction. There were 10,000 such plants by 1897. The age of electricity had arrived.

Motion Pictures

In the late 1880s, Edison turned his attention to what would be his last great series of inventions — the basic apparatus for motion pictures:

Source: Dagobert D. Runes, ed., *The Diary and Sundry Observations of Thomas Alva Edison* (NY: The Philosophical Library, Inc., 1948).

What primarily interested me in motion pictures [was]

the hope of developing something that would do for the eye what the phonograph did for the ear. That was the broad purpose, but how to accomplish that purpose was a problem which seemed more impossible the longer I studied it. It was in 1887 that I began my investigations.... The experiments in a laboratory consist mostly of finding that something won't work. The worst of it is you never know beforehand and sometimes it takes months, even years, before you discover you had been wrong all the time....

Edison finally succeeded in developing a workable motion picture camera, the kinetograph.

[Our first pictures] were shown...in an apparatus we christened "The Kinetescope," consisting of a cabinet equipped with an electrical motor and battery, and carrying a fifty-foot band of film, passed through the field of a magnifying glass. They attracted quite a lot of attention at the World's Fair in Chicago in 1893....

Edison's kinetoscope was a great success for several years. People stood in line outside "kinetoscope parlors" for a chance to watch through the "peep-hole" viewer. But, as in the case of the phonograph, Edison failed to see the potential of motion pictures. He decided against developing a projector and movie screen, saying that too many people could see a film at the same time and they would soon lose interest. Others developed the apparatus for the projector-and-screen films and, by the early 1900s, the modern motion picture industry was underway. Edison again quickly re-entered the field and soon became a powerful figure in silent movies, in partnership with our next inventor, George Eastman.

George Eastman and His Kodak

When you see photographs of people taken before 1880, you might notice that no one is ever smiling. The reason for this is that the subject had to hold the pose for many seconds, and sometimes even for minutes, while the picture was being taken. That was only one of the problems of photography before George Eastman (1854-1932) began his inventing career.

Eastman was a bank clerk in Rochester, NY, and an amateur photographer. In 1878, he began working on ways to improve the camera, which required heavy and fragile glass plates, along with a wagon-load of chemicals and equipment. By the early 1880s, he found that a new product — celluloid — could replace the glass plates and could be formed into a compact roll. Eastman developed a small box camera with roll film and was ready to go into business.

He wanted a simple, unique name for his camera, one that would be recognized and remembered in any language. He describes how he came up with the name:

Source: C.W. Ackerman, *George Eastman* (Boston: Houghton Mifflin Co., 1930).

I devised the name myself....The letter "K" had been a favorite with me — it seems a strong, incisive letter, and my mother's maiden name began with it....It became a question of trying out a great number of combinations of letters that made words starting and ending with "K." The word "Kodak" is the result.

THE

KODAK

AT THE

North Pole.

~~~~~~

2,000 Pictures

MADE BY

Lieut. Peary

Among

Greenland's

Icy

Mountains.

The Explorer Endorses the Kodak.

"My pictures were 'all taken with a Kodak and I regard the Kodak as responsible for my having obtained a series of pictures which in quality and quantity exceed any that have been brought back from Greenland and the Smith Sound region."

R. E. PEARY, U. S N.

~~~~~~~

EASTMAN KODAK CO.,

Send for Kodak Catalogue.   **Rochester, N. Y.**

An early ad for Kodak carried a testimonial from Lt. Peary

The cameras sold for twenty-five dollars and came with a 100-exposure roll of film. When the roll was exposed, the photographer returned the entire camera to the Kodak Company in Rochester for developing and a new roll of film ($10 fee).

Eastman proved to be a masterful businessman as well as a brilliant inventor. He continued to improve the camera and lower the price. His "Brownie" camera, developed in the 1890s, sold for a dollar and remained popular for half a century. And his slogan — "You Press the Button, We Do the Rest" — became one of the most successful advertising phrases. As a contemporary jingle put it:

To diagnose

Our modest pose,

The Kodak does its best;

If evidence you would possess,

You only need a button press,

And we do all the rest.

— Eastman Kodak Company

The Kodak cameras, and the use of Kodak film in Edison's motion pictures, made Eastman a very wealthy man. In the late 1920s, he began giving away his fortune of $75 million, mostly to colleges and universities, and often anonymously as a gift from "Mr. Brown."

The First Automobiles

*"The ordinary `horseless carriage' will never,
of course, come into as common use as the bicycle."*

— *Literary Digest*, 1899

By the time this prediction was made in 1899, there were only about fifty automobiles in the entire country, and inventors had been tinkering with them for more than twenty years. A number of thorny problems had to be worked out, including how to make a small engine with a portable and efficient fuel supply and devising systems for starting, stopping, and steering the vehicle. Following ideas developed in Europe, the Duryea brothers of Springfield, MA, developed what is regarded as the first American automobile in 1891, and later actually built twelve more. Not far behind was one of Edison's employees working in Detroit, a young mechanic named Henry Ford (1863-1947). By 1893, Ford had designed and built his "Quadricycle."

The Quadricycle

Source: Henry Ford, *My Life and Work* (New York: Doubleday, Page & Co., 1922), pp. 32-33.

[My] first car had something of the appearance of a buggy. ... The power was transmitted from the motor to the counter-shaft by a belt and from the countershaft to the rear wheel by a chain.... There were two speeds — one of ten and the other of twenty miles per hour — obtained by shifting the belt, which was done by a clutch lever in front of the driving seat.... There was no reverse.... The wheels were 28-inch bicycle wheels with rubber tires....

My "gasoline buggy" was the first and for a long time the only automobile in Detroit. It was considered to be something of a nuisance, for it made a racket and scared horses. Also it blocked [horse-drawn] traffic. For if I stopped my machine anywhere in town a crowd was around it before I could start it up again. If I left it alone even for a minute some inquisitive person always tried to run it. Finally, I had to carry a chain and chain it to a lamp post whenever I left it anywhere.

It was not until 1908 that Ford worked out a system for producing automobiles. By combining Whitney's system of interchangeable parts with Colt's assembly line, he was able to speed up production and lower costs. It was Ford's innovations that popularized the automobile by making it affordable. Ford's Model T stayed in production for twenty years and more than 15 million of the "Tin Lizzies" were sold — one out of every three cars on the road.

Safety, Comfort, Convenience

Every major invention of the 19th century — the telegraph, telephone, railroad, electric lights, and others — led to the creation of entire new industries and to new inventions. Just as the telegraph led Bell to invent the telephone, and the telephone led Edison to the idea of the phonograph, other inventors found their imagination spurred by the latest developments.

Here is a brief Honor Roll of some of the inventions that grew out of other inventions:

The Sholes Typewriter

Christopher Latham Sholes (1819-1890), a printer in Wisconsin, was trying to develop a printing device for telegraph messages. This led him, in stages, to invent the typewriter, with a keyboard that essentially remains the same on today's typewriters and computers. He wrote to a friend:

Source: G.T. Richards, *The History and Development of the Typewriter* (London: HMSO, 1964).

Its simplicity cannot be equalled, it being more simple than a piano, and so less liable to get out of order. A child may thump its keys with pleasure and do it no harm. It works also as easily as a piano....I am inclined to think that in a day's writing the use of the pen would be found much more tiresome than the use of this machine. As to rapidity, I am working this about as fast as ordinary writing....

Lillian Sholes demonstrates the typewriter invented by her father.

Sholes sold his patent to the Remington Arms Company in 1873. But it took another twenty years for people to realize the value of the machine for both business and personal use.

George M. Pullman and Railroad Comfort

In the mid-1800s, railroad travel was fast and convenient, but also uncomfortable. Sleeping accommodations were primitive. For meals, the passengers waited for a station stop, then rushed into the station restaurant to wolf down a meal before the signal to re-board.

George M. Pullman (1831-1899) changed all that. He transformed railroad cars into "moving hotels." The Pullman Sleepers had comfortable beds and a daily change of linen. His dining cars became famous. One obviously satisfied diner wrote:

Source: W.F. Rae, *Westward by Rail* (London: Isbister, 1870, reprinted by Arno Press, New York, 1973).

The choice [of foods] is by no means small. Five different kinds of bread, four sorts of cold meat, six hot dishes, to say nothing of eggs cooked in seven different ways and all the seasonable vegetables and fruits, form a variety from which the most dainty eater might easily find something to tickle his palate and the ravenous to satisfy his appetite....To breakfast, dine, and sup in this style while the train is speeding at the rate of nearly thirty miles an hour is a sensation of which the novelty is not greater than the comfort.

British author Bruce Norman describes the social significance of Pullman's achievement:

Source: Bruce Norman, *op. cit.,* p. 75.

...The Pullman cars were mass-produced. And, for the first time, mass-production could mean mass luxury. The American System of Manufacture responded successfully to the new demand and showed that it could not only produce fast and cheap but could put a really high surface gloss on the basic technology. That it could achieve by machinery what could previously only have been done well by hand.

George Westinghouse: Railroad Safety

After serving in both the Union Army and Navy in the Civil War, young George Westinghouse (1846-1914) invented an air brake for trains in 1868. The only way to stop a train until then was for a brakeman stationed between each car to turn a wheel that slowly applied a brake. Westinghouse devised a system with a compressed air pipe running the length of the train. When the engineer threw a switch cutting off the air, brakes were automatically applied to the wheels of every car.

Westinghouse had trouble selling his system to the railroads, until one company agreed to a test. An engineer named Tate got the train's speed up to thirty miles an hour, when he suddenly saw a horse-drawn wagon on the tracks:

Source: *Popular Science Monthly*, Sept., 1927.

Tate grasped the brake control and twisted desperately. With a mighty lurch, the train stopped dead.

Picking themselves up from the floor, the passengers in the rear car scrambled to the platforms and sprang to the ground. They found Tate assisting the terror-stricken [wagon driver] to arise — four feet from the train's cowcatcher.

So bruised and ruffled were the witnesses that the significance of the event was slow in dawning. And then they comprehended. In saving a human life, the air brake had demonstrated its own efficiency.

The air brake rapidly became standard on all trains and Westinghouse went on to more inventions, receiving more than 400 patents. He formed Westinghouse Electric Company in 1886 and engaged in a long battle with Edison over electrical current. Edison stubbornly insisted on using direct current (DC), but Westinghouse eventually proved that alternating current (AC) was safer and more economical.

Putting Electricity to Work

Frank Julian Sprague (1857-1934) has all but disappeared from history books, but he played a key role in demonstrating the uses of electrical power. As a young assistant of Edison, Sprague became fascinated with the potential of electric motors. In 1884, he formed the Sprague Electric Railway and Motor Company. After perfecting his motors, he built the first electric-trolley system in 1888 in Richmond, Virginia. More than 100 other cities soon adopted this early form of mass transit, and Sprague sold his business to Edison's General Electric Company.

Sprague next developed an electric motor for elevators and sold that business to the Otis Elevator Company. His work contributed to both elevated trains and subways. After 1900, he demonstrated ways to use electric motors to power small tools and household appliances. Edison created the system of generating and distributing electrical power, but Sprague deserves much of the credit for showing how that power could be used.

Inventions Great and Small

Nineteenth-century inventors came up with a wide array of devices to meet every human need they could imagine. Most of these, fortunately, never went into production, but they did receive patents.

The patent for the 1868 "Improved Burial-Case" promised:

Source: *The Smithsonian Book of Invention* (NY: W. W. Norton, 1978).

"Should a person be interred ere life is extinct, he can, on recovery to consciousness, ascend from the grave and the coffin by ladder; or, if not able to ascend by ladder, ring the bell, thereby giving an alarm."

New technology brought with it the need for newly trained workers. Rose Pastor Stokes, who had done piece work by hand for many years, recorded the benefits from learning to roll cigars by machine:

Source: Rose Pastor Stokes, "I Belong to the Working Class" (Around 1892) From Robert D. Marcus and David Burner, *America Firsthand*, Vol II, (New York: St. Martin's Press, 1995), p. 74.

With the monopoly of the newly-invented suction machine, by which a worker could turn out many times the number of cigars made by skilled hand labor, the Cigar Trust came to existence. It was spreading westward from New York. It needed workers to operate the new machines. A Mr. Young, foreman at Baer's had shown off my skill, economy of motion, and economical use of material, to visiting buyers and "manufacturers." Mr. Young it was who was now engaged by the trust to start their Cleveland factory, and who, in turn, engaged me to learn the suction-machine method, and to teach it to other workers. Soon I was given charge of an entire floor, at fifteen dollars a week.

The Uninvited

This book's survey of 19th century inventors and their inventions has only scratched the surface. There are countless others who made extraordinary contributions, great and small. Walter Hunt (1796-1859), for example invented dozens of items, including the safety pin, which he devised in three hours in order to pay a debt. More significant was Ottmar Merganthaler (1854-1899) whose Linotype machine made high-speed printing possible.

Another of the heroes was a man with the ambitious name of Thaddeus Sobieski Coulincourt Lowe (1832-1913). During the Civil War, Professor Lowe became the nation's first "aeronaut," ascending in a balloon, equipped with a telegraph, to scout Confederate positions for the Union Army. The invention of the airplane, however, would have to wait for the Wright Brothers in 1903.

While men like Lowe, Mergenthaler, and Hunt are scarcely remembered today, they at least enjoyed considerable fame and rewards during their lifetimes. The same cannot be said for two groups of inventors we'll call "the uninvited," since, whatever their achievements, they were not going to receive much credit in a society dominated by white males.

The largest group of the uninvited, of course, are women. Because men kept the records and wrote the stories, we may never know how much women contributed to the amazing century of invention. There is some evidence, for example, to suggest that Catherine Greene helped Eli Whitney invent the cotton gin, and that Cyrus McCormick's wife contributed key ideas to the invention of the reaper. There is no way to verify such suspicions, since the men secured the patents and gained the fame.

Women Inventors?

Women in the 19th century had little opportunity to be inventive. The accepted role of women as homemakers was rarely challenged, except by a handful of daring pioneers. And the task of homemaking in the days before labor-saving devices required a huge amount of time and energy. In addition, working in the home gave women little chance to gain skills in mechanics or in the use of power sources like electricity and steam. In addition, laws in many states prohibited women from owning property or signing documents, such as patent applications.

Even with these obstacles, a few determined women defied convention and entered the world of invention. Some examples:

In 1809, Mary Dixon Kies became the first woman to receive a patent — for a process of using thread in the weaving of straw hats.

During the Civil War, Martha Coston perfected flares for night use on ships and received her patent in 1871. The system was used for many years.

Helen Augusta Blanchard had more than twenty patents to her credit, dealing with improvements in sewing machines. Her ideas earned royalties from the American Sewing Machine Company.

The most prolific woman inventor of the time was probably Margaret E. Knight (1838-1914). She invented the machinery for manufacturing the square-bottomed paper shopping bag, then spent several years in court defending her patent. Among her other inventions was the Knight Silent Motor, a gasoline engine for automobiles. She is estimated to have made nearly 100 inventions, but she usually assigned the patent rights to friends or financial backers. She was known in feminist circles as "the feminine Edison."

Machine for manufacturing a square-bottom paper bag, perfected by Margaret E. Knight

A product of Madam C. J. Walker, a successful, black female entrepreneur, who not only developed beauty products for black women, but ran training schools to show sales people and customers how to use the products.

African-American Inventors

African Americans in the 19th century, women as well as men, had even less opportunity than white women to develop inventions or to benefit from them. As in the case of women, a number made their way through prejudice and the laws to develop inventions. A study completed by the Patent Office in 1913 found more than 1,000 patents granted to African-American men and women. Some examples:

Thomas L. Jennings received the first known patent by an African American in 1821 for inventing the first dry cleaning machine for clothing.

Norbert Rillieux (1806-1894) developed a device for refining sugar using a vacuum-evaporation system which, in modified form, is still in use. His invention was also adapted by Gail Borden for condensing milk.

Lewis Howard Latimer (1848-1928) worked with Thomas Edison for many years and was credited with a number of inventons, including an improved filament for light bulbs and an improved socket. Although he received little acclaim during his lifetime, when he died in 1928 the *Edison Pioneers* (members of the early group who had worked with Edison) offered these words on Latimer:

Source: William Loren Katz, *Eyewitness: A Living Documentary of the African American Contribution to American History,* New York: Simon & Schuster, Touchstone Edition. 1995.

... In this office he became interested in draughting and gradually perfected himself to such a degree as to become their chief draughtsman.... It was Mr. Latimer who executed the drawings and assisted in preparing the applications for the telephone patents of Alexander Graham Bell. In 1880 he entered the employ of Hiram S. Maxim, Electrician of the United States Electric Lighting Co., then located at Bridgeport, Connecticut. It was while in this employ that Mr. Latimer successfully produced a method of making carbon filaments for the Maxim electric incandescent lamp, which he patented. His keen perception of the possibilities of the electric light and kindred industries resulted in his being the author of several other inventions.... In 1884 he became associated with the Engineering Department of the Edison Electric Light Company.

He was of the colored race, the only one in our organization, and was one of those to respond to the initial call that led to the formation of the Edison Pioneers, January 24,1918. Broadmindedness, versatility in the accomplishment of things intellectual and cultural, a linguist, a devoted husband and father, all were characteristic of him, and his genial presence will be missed from our gatherings.

In 1872, Elijah McCoy invented a device that continually oiled the parts of a railroad engine, even while the train was in motion. The invention was soon widely adopted. It was also widely imitated, which may have been the reason for the term "the real McCoy."

Jan E. Matzeliger (1852-1889) was probably the most influential African-American inventor of the century, although he never received much notariety or financial success. Beginning in 1877, he spent six years perfecting a shoe-making machine. His machine revolutionized the shoe industry by making mass production possible, and was responsible for the multimillion dollar growth of the United Shoe Company, which bought his invention for a pittance. The editor of a Lynn, Massachusetts newspaper described Matzeliger as "a man not only of wonderful mechanical ability, but a man of equally wonderful energy and tenacity of purpose." (Source: Katz, *ibid.*)

George H. Murray, a former slave, was elected to Congress in 1892. He addressed the House of Representatives about the contributions of blacks in the area of invention.

Source: Congressional Record, 53rd Congress, 2nd session, 8382, found in Katz, *ibid.*

I hold in my hand a statement prepared by one of the assistants in the Patent Office, showing the inventions that have been made by colored men within the past few years....

This statement shows that colored men have taken out patents upon almost everything, from a cooking stove to a locomotive. Patents have been granted to colored men for inventions and improvements in the workshop, on the farm, in the factory, on the railroad, in the mine, in almost every department of labor, and some of the most important improvements that go to make up that great motive power of modern industrial machinery, the steam engine, have been produced by colored men....

... Mr. Speaker, the colored people of this country want an opportunity to show that the progress, that the civilization which is now admired the world over, that the civilization which is now leading the world, that the civilization which all the nations of the world look up to and imitate— the colored people, I say, want an opportunity to show that they, too, are part and parcel of that great civilization....

Mr. Speaker, in conclusion I ask the liberty [of] appending to my remarks the statistics [a list of 92 patents] to which I referred.

There was no objection.

Suggested Further Reading

Haskins, James. Outward Dreams: *Black Inventors and their Inventions*, New York: Walker & Company, 1991.

James, Portia P. *The Real McCoy: African-American Invention and Innovation*, Washington, D.C.: Smithsonian Institution Press, 1989.

King, David C. *Thomas Alva Edison: The King of Inventors*, Carlisle, MA: Discovery Enterprises, Ltd., 1995.

Lafferty, Peter. *The Inventor Through History*, New York: Thompson Learning, 1993.

Norman, Bruce. *The Inventing of America*, New York: Taplinger Publishing Company, 1976.

Richardson, Robert O. *The Weird and Wondrous World of Invention*, New York: Sterling Publishing Co., 1990.

Showell, Ellen H. and Amram, Fred M. B. *From Indian Corn to Outer Space: Women Invent America.* Peterborough, NH: Cobblestone Publishing, Inc. 1995. (Young adult)

Stanley, Autumn. *Mothers and Daughters of Invention: Notes for a Revised History of Technology*, Metuchen, NJ: Scarecrow Press, 1993.

About the Editor

David C. King compiled the readings for this volume and wrote the portions that are not firsthand accounts. He has written more than thirty books for young people, primarily in American history, including *Thomas Alva Edison: The King of Inventors* and *First Facts About American Heroes* which was selected by the Children's Book Council as a Notable Book for 1997. He and his wife Sharon Flitterman-King live in Hillsdale, New York.